Portage Public Library

DISCARDED

COCKROACHES
CREEPY CRAWLERS

Lynn Stone

The Rourke Book Co., Inc.
Vero Beach, Florida 32964

© 1995 The Rourke Book Co., Inc.

All rights reserved. No part of this book may be reproduced or utilized in any form or by any means, electronic or mechanical including photocopying, recording or by any information storage and retrieval system without permission in writing from the publisher.

PHOTO CREDITS
Cover, pages 4, 12 and 18 © James P. Rowan; title page, pages 7, 8, 13, 15, 17 © Lynn M. Stone; page 10 © James H. Robinson; page 21 © Breck P. Kent

Library of Congress Cataloging-in-Publication Data

Stone, Lynn M.
   Cockroaches / by Lynn Stone.
     p. cm. — (Creepy crawlers)
   Includes index.
   Summary: Describes what cockroaches look like, what they eat, where and how they live, who their relatives are, and how their antennas serve as senses.
     ISBN 1-55916-161-2
     1. Cockroaches—Juvenile literature.   [1. Cockroaches.]
I. Title II. Series: Stone, Lynn M. Creepy crawlers
QL 505.5.S76 1995
595.7'22—dc20   95–16555
                                                              CIP
                                                             AC

Printed in the USA

TABLE OF CONTENTS

Cockroaches	5
What Cockroaches Look Like	6
Kinds of Cockroaches	9
What They Eat	11
Where Cockroaches Live	14
Relatives	16
How Cockroaches Live	19
A Cockroach's Senses	20
Cockroaches and People	22
Glossary	23
Index	24

COCKROACHES

Other than the scientists who study them, perhaps no one loves cockroaches. People usually greet these insects with a look of disgust or a loud "Yuk!"

Cockroaches can be nasty house guests. They can be smelly and spread disease. Most kinds of cockroaches, however, live in wild places and have nothing to do with people.

This Costa Rican cockroach on the forest floor is not a threat to someone's home

WHAT COCKROACHES LOOK LIKE

Cockroaches, like other insects, have no bones. They do have a hardened, flat back, however, that serves them as a skeleton of sorts.

Cockroaches have six legs in three pairs to move their body along. Some cockroach bodies are no larger than ants. Others are about the length of a baby's first finger. Really big cockroaches are about the size of a sparrow.

Briefly on its back, a cockroach shows its six insect legs

KINDS OF COCKROACHES

Cockroaches are not all alike. Scientists have named about 4,000 **species** (SPEE sheez), or kinds. Many more cockroach species, waiting to be "discovered," are hidden away in the world's rain forests.

Most cockroaches are brown, but you don't have to be dull to be a cockroach. One species, for example, is bright green. Another has a **translucent** (tranz LOO sent) body, which means it's almost as clear as water.

Cockroaches can be quite colorful

WHAT THEY EAT

The cockroach family has been around for perhaps 350 million years. One reason for their success in surviving is their diet. Unlike many other animals, cockroaches can—and do—eat almost anything.

Their little mouths crunch up dead insects, bread crumbs, and plant material. They even gobble up their old outer "skeleton" when a new one replaces it.

When a cockroach egg doesn't hatch, it's not wasted. The egg layer is likely to eat it!

Cockroaches eat all kinds of things, including leftovers in the kitchen

While molting, a cockroach sheds its old skin (right)

Hard to find among the twigs, a walkingstick is a cousin of the cockroach

WHERE COCKROACHES LIVE

You may have cockroaches in your home. They don't like light, so you'll rarely see them during the day.

The great majority of cockroaches, however, live in warm, wet jungles, the world's rain forests. Others live in hot desert areas, woodlands, and caves.

Cockroaches will move in with almost anyone—human or animal. They share burrows with tortoises and nests with birds and bees.

This cockroach is at home in Texas leaf litter

RELATIVES

Cockroaches are **invertebrates** (in VERT uh brayts) as well as insects. Invertebrates are all the animals without backbones—snails, spiders, worms, insects, and other groups.

The closest relatives of cockroaches are certain other insects, the crickets, grasshoppers, walkingsticks, katydids, and mantids.

The cockroaches and their close cousins have long, whiplike **antennas** (an TEN uhz) and wings. These insects seem to prefer walking to flying, however.

This bright green katydid is a cousin of the cockroaches

HOW COCKROACHES LIVE

The cockroaches that live in people's homes are truly **nocturnal** (nahk TUR nul)—creatures of the night. They hide in cracks and other dark, moist places during the day. At night they crawl out to hunt for food. They love crumbs and table scraps. If someone turns on a light, the cockroaches run like rabbits.

Wood cockroaches live outdoors, under stones and logs.

Some cockroaches hatch from eggs. Others are born alive.

A wood cockroach in Illinois crawls along with its egg case

A COCKROACH'S SENSES

A cockroach doesn't have a nose to smell or a tongue to taste. It does have a pair of antennas, however, that help it taste, smell, and feel air movement.

In addition, cockroach heads have tiny, fingerlike parts called **palps** (PALPS). Sensitive hairs on the palps taste food. If the cockroach likes the taste, the palps deliver it to the mouthparts.

A hissing roach of Madagascar shows its fearsome face

COCKROACHES AND PEOPLE

People with cockroaches in their homes declared war on these insects long ago. Poison bait is the main weapon, and it kills millions of cockroaches. Then millions more take their place.

Many people are allergic to cockroaches. Cockroaches can make human eyes water. The insects can also cause breathing problems, especially in people with **asthma** (AZ muh). People with asthma sometimes suffer from shortness of breath. Cockroaches, dead or alive, can make their breathing even more difficult.

Glossary

antennas (an TEN uhz) — on the heads of cockroaches and certain other animals, stalklike structures used to sense smell, movement, and location

asthma (AZ muh) — a condition that causes difficulty in breathing

invertebrates (in VERT uh brayts) — the simple, boneless animals, such as insects, worms, snails, starfish and slugs

nocturnal (nahk TUR nul) — active at night

palps (PALPS) — a mouthpart of certain insects, such as a cockroach

species (SPEE sheez) — within a group of closely related animals, one certain kind, such as a *mallard* duck

translucent (tranz LOO sent) — partly clear, like thick, dull glass

INDEX

antennas   16, 20
asthma   22
crickets   16
disease   5
egg   11, 19
grasshoppers   16
homes (human)   14, 19, 22
insects   5, 6, 16
invertebrates   16
katydids   16
legs   6
mouthparts   20
mouths   11
nocturnal   19
palps   20
people   5, 22
poison bait   22
rain forests   9, 14
skeleton   6, 11
snails   16
species   9
spiders   16
walkingsticks   16
worms   16